The Controversial Coyote

by the same author
DINOSAURS AND THEIR WORLD

The Controversial Coyote

PREDATION, POLITICS, AND ECOLOGY

BY LAURENCE PRINGLE ✳ ✳ ✳ ✳ ✳

Illustrated with photographs ———————————————————

Harcourt   Brace   Jovanovich   *New York and London*

Printed in the United States of America

Title page photograph: *Guy Connolly*

Library of Congress Cataloging in Publication Data

Pringle, Laurence P
The controversial coyote.

Bibliography: p.
Includes index.
SUMMARY: Discusses controversial attempts to control
the coyote population of the United States.
1. Coyotes—Control—United States—Juvenile
literature. 2. Coyotes—United States—Juvenile
literature. 3. Sheep—United States—Juvenile
literature. 4. Predation (Biology)—Juvenile
literature. [1. Coyotes. 2. Predatory animals]
I. Title.
SB994.C8P73   639'.97'974442   76-46789
ISBN 0-15-219930-6

First edition

B C D E F G H I J K

Contents

About This Book

The coyote is the most controversial wild animal in North America and seems destined to hold that honor for a long time. No other animal has stirred up more human emotion, aroused more political furor, or cost so much money in efforts aimed at its control.

The coyote does this by simply being the coyote—a medium-sized predator with forty-two teeth, trying to eat, raise some young, keep alive. Sometimes it kills sheep, goats, calves, and other small livestock. It can cause serious economic loss to ranchers and is the main target of a federally administered predator control program now more than sixty years old.

Within the past two decades this campaign has faced increasing criticism. The changes made so far have not been satisfactory to either side of the controversy, so it rages on. Sheepmen and environmentalists argue and contradict each other. Which can you believe?

Perhaps neither. Nonsense and half-truths abound. Some sheep ranchers exaggerate their losses to predators and feel that the West won't be properly tamed until all coyotes are dead. Many people who oppose predator control are poorly informed about coyotes and also about the effects of the predator-control program. A photograph of dead coyotes hung on a fence may arouse lots of emotion, but it reveals nothing about the actual status of a coyote population, which may be thriving.

Whom can you believe? I suggest a third group: the wildlife biologists who are studying (at last) the complex relationships between coyotes and their environment, including sheep. Their work has already helped explode some myths about the lives of coyotes and the effects of predator control. In time they will help shed more light on a subject where mostly heat existed before.

Of course, wildlife biologists are human too; their feelings may influence their objectivity. Those I have met have considerable affection and admiration for coyotes but try not to let these feelings interfere with the facts that unfold. Their research and conclusions are open to close scrutiny. Unlike many people who hold strong opinions about the coyote-sheep controversy, these professional biologists collect data and make observations "in the field," from populations of real coyotes and real sheep. And they do not expect simple solutions to complex problems.

This book emphasizes the discoveries of these scientists, stresses the need for patience to allow them to find solutions to the coyote-sheep problem, and is dedicated to them.

Laurence Pringle

The Controversial Coyote

ONE

The
Rising Storm

Bloodthirsty, savage killers. Vile murderers. Predators. They're all one and the same in the minds of many people. Although predators simply kill other animals in order to eat, for most of human history people have thought of them as "evil" enemies, as animals of little or no value. "The only good predator is a dead one" was the persistent philosophy.

The word predator usually brings to mind pictures of attacking lions or wolves or hawks, but such creatures as toads, robins, and ladybugs are also predators. No one complains about "vicious robins" killing earthworms. In reality, humans themselves are the master predators of the earth. So when people call predators "bad," they are usually referring to certain ones that may do them some real or imagined harm—the infamous "big bad wolf," a weasel in the henhouse, a coyote that kills sheep. Some hunters even consider themselves in competition with foxes and other predators. They believe that any rabbit or other game killed by a predatory animal is a personal loss. In fact, another definition of a predator is "any critter that gets something you wanted before you do."

Since most people know very little about predators,

3

The coyote—an "evil killer" or a valuable, misunderstood mammal?
This is a tame two-year-old coyote.   (*Guy Connolly*)

they often have strange notions about them. Some of these
ideas come directly from fairy tales and stories like "Little
Red Riding Hood." Many people still believe that wolves
attacked early settlers in North America, yet there is not
one proved case of this happening, except in the rare instances when a wolf had the disease called rabies.

Predators—especially large ones such as wolves—evoke
feelings of fear and awe. In their imaginations, people credit
predators with all sorts of powers and deeds. No predators

in real life are as fierce, cunning, and strong as they are in people's fantasies. By imagining that predators are cruel, dangerous, and powerful, people feel that they accomplish something good and important when they kill them.

But as we learn more about predators, an increasing number of people are beginning to believe that predators have value. This belief is not new. In the past, so-called primitive people respected the predators that lived around them. These people were in close touch with the land and its life. North American Indians seemed to understand that wolves, bears, and other predators were an important part of nature. A Sioux chief said, "Great contentment comes with the feeling of friendship and kinship with the living things about you. The white man seems to look upon all animal life as enemies, while we look upon them as friends and benefactors."

The Indians and their beliefs were swept aside, however, by colonists from Europe. These settlers had good reason to shoot and trap some predators. The loss of a cow or pig brought real hardship, so the death of a bear or wolf was good news. Besides, the pelt could be used to make clothing or traded for food, gunpowder, or some small luxury.

But settlers weren't content to kill only the predators that directly threatened their livestock. They had a continent to "tame" and believed that part of that task was wiping out all "evil" predators. Bounties were paid for wolves as early as 1630, and most of United States history is marked by an undeclared war on predators.

Actually, for many years not only predators but all of nature was under attack. Forests and other wildlife habitats were ruthlessly cut and burned. Deer, wild turkeys, beavers, and bison were hunted or trapped to the edge of extinction in many areas. And some species, such as the passenger pigeon, were completely wiped out.

By the late 1800s, some people had become alarmed at the wholesale destruction of the land and its life. They fought for laws that protected deer and other game animals from market hunting. They helped bring about the creation of wildlife refuges, as well as state and national forests. Many of these early conservationists, as they came to be called, were hunters of deer and other game animals who were mainly concerned about the future of their sport. Theodore ("Teddy") Roosevelt is probably the best known. No one questions the great good he did in helping to create national forests and national parks. But his attitude toward predators was just as uninformed and simplistic as that of most of the people of his time. Describing a treed cougar he was about to shoot, he once wrote," . . . the big horse-killing cat, the destroyer of the deer, the lord of stealthy murder, facing his doom with a heart both craven and cruel."

This was written in 1913. Hunters still believed that predators were competitors whose death would guarantee more game for sportsmen. The emphasis was on conservation of game, not of all wildlife. So the war on predators continued. In the eastern United States the larger species had already been wiped out, but wolves, coyotes, cougars, and bears were still abundant in the West. They preyed upon deer and other wild game, as they always had. Sometimes they also killed livestock that was raised on ranches and on the vast acreages of federal land where grazing was allowed—land that had formerly been their hunting grounds.

In 1915 the United States Congress gave funds to the Bureau of Biological Survey so that it could join the war

Wolves are not the "super-predators" that people imagine. Studies of wild wolves reveal that they often fail in attempts to kill healthy prey and sometimes go many days without food.   (*Laurence Pringle*)

Persecution of the cougar, or mountain lion, has caused it to disappear from most of its former range.　(*U.S. Fish and Wildlife Service*)

on predators. These funds were supplemented by contributions from several Western states and also from individual stock raisers. More than three hundred hunter-trappers were hired. This was just a beginning; the anti-predator effort grew with the years.

New Mexico had one of the most vigorous predator-killing programs. A man named Aldo Leopold, employed by the United States Forest Service, was that state's chief advocate of predator extermination. In 1920 Leopold proudly reported to the National Game Conference in New York City that New Mexico had reduced its wolf population from three hundred to thirty in just three years. "It is going to take patience and money to catch the last wolf or lion in New Mexico," he said, "but the last one must be caught before the job can be called fully successful."

Leopold currently is considered the "father" of the profession of wildlife management, which seeks to restore and maintain wildlife populations. At first he didn't consider predators worth saving; they were "varmints" that infested the land. Leopold had an open mind, however, and later events caused a dramatic change in his attitudes toward predators and all of nature.

Deer were the animals that helped change Leopold's mind. In 1927 there were startling reports about deer populations in Black Canyon, part of the Gila National Forest in New Mexico. The deer were so plentiful that they had damaged the shrubs and other plants upon which they fed. Most of the deer seemed to be in poor physical condition.

The same situation had also been reported from the Kaibab National Forest in Arizona. Thousands of deer starved to death there each year. According to one report, "Those that lived ate every leaf and twig until the whole country looked as though a swarm of locusts had swept through it, leaving the range (except for the taller shrubs and trees) torn, gray, stripped, and dying." Deer population "explosions" were also occurring in parts of other states, including Pennsylvania. A startling fact emerged: it was possible to have *too many* deer.

Evidence from the Kaibab showed that the removal of predators was partly responsible for the explosion of deer numbers there. The Kaibab was the most famous case, but the same conditions developed in other regions where deer, elk, and other large plant eaters were protected and their predators wiped out. Before long, Aldo Leopold and others were urging that cougars and other predators

An overabundance of deer, with resulting damage to plant life, is still a problem in parts of several states.

(*Wisconsin Conservation Department*)

be protected in national parks and other refuges where they were the main check on populations of big game.

At this point—in the late 1920s—our understanding of predators and of nature in general was still poor. In 1933 Aldo Leopold wrote a book about game management. In it he described a game manager as a person who plays a game of chess with nature but who can just barely see the board, the pieces, and the rules.

OUR GAME WITH NATURE

Our chances in this game improved as the science of ecology developed. The term "ecology" was coined in 1866 by the

German biologist Ernst Haeckel. He defined it as "the whole science of the relations of the organism to the environment." He meant environment in its broadest sense, including soils and climate as well as living things. However, this idea was neglected for decades. When biologists did begin to study communities of living things, they did not at first consider all factors in the environment. Botanists studied communities of plants and noticed that they changed over the years. Zoologists noticed that animal communities changed along with the plants. For a time no one considered the interrelationships between plants and animals, or among other factors in the environment.

By the mid-1930s, however, some biologists were exploring the concept of ecology in its broadest sense. They were beginning to think of the land itself as an organism, with living and nonliving parts, all interrelated and interdependent. For Aldo Leopold, this concept took hold in 1936, on a pack trip to the wilds of northern Mexico. Deer were plentiful. So were wolves and cougars. They thrived together, and so did other animals and plants. Leopold realized that he was seeing ecologically healthy land for the first time.

As the years passed, research revealed more and more of the board, pieces, and rules of the human chess game with nature. The emerging science of ecology brought an end to some long-cherished ideas. Labels like "good" and "bad" became irrelevant. Leopold wrote, "The last word in ignorance is the man who says of an animal or plant, 'What good is it?' If the land mechanism as a whole is good, then every part is good, whether we understand it or not."

By the 1960s the rapidly developing science of ecology had begun to change the attitudes of many people. Ecology has been called the first ethical science. An understanding of ecology evokes respect for all of nature. Increasing numbers of people believe that they are part of nature and that the old drive to "conquer" nature can only lead to self-defeat. The goal of our game with nature *is not to win, but to keep on playing.*

The development of this philosophy caused some changes in government policies. For example, some states have passed laws that protect surviving populations of predators. The profession of game management has broadened into wildlife management, concerned with maintaining and restoring entire plant-animal communities rather than just huntable species.

However, these signs of an emerging ecological conscience took decades to develop. All the while, the war against predators continued, especially in the West. A Division of Predator and Rodent Control was established in the Department of the Interior. Its annual budget grew steadily, reaching eight million dollars in the 1960s. Wolves and grizzly bears were wiped out in state after state; the range of cougars was greatly reduced. Yet the predator-control program continued, the budget grew, the war went on, mostly because of one animal—the coyote.

Sometimes called "brush wolf," the coyote was once just one of several large predators in the West. (Its name can be pronounced *ky-oat* or *ky-o-tee.*) It was most abundant on open prairies and deserts. Coyotes were scarce or nonexistent wherever wolves were plentiful. The extermination of wolves allowed coyotes to expand their range. Other changes in the environment also favored the coyote. Overgrazing by livestock damaged the lush rangeland grasses. Weeds and other plants that were unpalatable to livestock flourished and provided food and shelter for rabbits, mice,

The coyote eats a wide variety of animal and plant food, but jack-rabbits are a major part of its diet in the West.

gophers—mammals that are the basic food of coyotes. People unwittingly created ideal conditions for the increase and spread of the coyote all over the West, and beyond.

None of this could have occurred if the coyote had not been the remarkable beast that it is. This animal's cunning and adaptability were recognized long ago by Indians of the Great Plains. To be called "coyote-smart" was a fine compliment because Indians considered the coyote to be the smartest animal on earth. In Indian tales, the coyote was usually a powerful and friendly figure. Some tribes called it "God's dog." Early settlers, ranchers, and government trappers had other names for it. The coyote has been called "the king of beasts," "a dare to civilization," and "the predatory animal menace of North America."

The harshest words for coyotes come from sheep raisers. Many sheepmen claim that coyotes killing their sheep are their most serious problem; some ex-sheepmen

Sheep en route to their winter range on public lands in Utah.
(*U.S. Bureau of Land Management*)

say that coyotes drove them out of business.

The sheep industry is concentrated in seventeen Western states where, for modest fees, many sheep are allowed to graze on millions of acres of public lands. In 1942 more than fifty million head of sheep were raised in North America. Although less than a third of that number are

now raised, sheep raising is still a major business in the West, and thus the sheepmen still have great influence on local and state governments and on the federal government in Washington, D.C. Political pressure and financial support from sheep ranchers have helped keep the war on predators under way.

MORE AND BETTER WEAPONS

At first the main weapons were guns, traps, snares, and the poison strychnine, which was put in chunks of animal fat or in carcasses on which coyotes might feed. Coyote dens were also sought; when these were found, the young were lured out or dug out and killed. Eventually more modern weapons, especially other poisons, were used. By the 1960s, Western coyotes were being hunted from low-flying airplanes and from snowmobiles. But the most effective weapons were poisons. They were used by several hundred government agents and also by ranchers.

One poisoning device was called the "coyote getter." To set this, first a hollow metal stake was driven into the ground. Then a firing mechanism and an explosive cartridge containing sodium cyanide were placed in the top of the stake. The "getter" was topped with a tuft of wool or other bait, and a food scent was smeared on it to make it more attractive.

Coyote getters were scarcely noticeable. Only the bait showed above ground, and it was about the size of a man's thumb. When a coyote found the bait and tugged at it, the cartridge exploded, shooting the cyanide into its mouth. The animal usually died within thirty to sixty seconds. Coyote getters were eventually replaced by similar devices called M-44s, which are powered by springs rather than by cartridges and are less hazardous to people.

The most popular poison was sodium monofluoracetate, usually called Compound 1080. Carcasses of sheep, horses, or other large mammals were injected with this poison, in hopes that coyotes would eat enough to kill them. The poison was most toxic to coyotes and other members of the dog family. One or two ounces of meat containing

Devices used to kill coyotes include steel leg-hold traps and the M-44, shown being installed in the ground. The latter is more likely to kill only coyotes and is less cruel because of its fast-acting poison.
(*Animal Damage Control Program, U.S. Fish and Wildlife Service*)

1080 was often enough to kill a coyote. Much larger amounts were needed to kill most of the other animals that might come along. Nevertheless, some other animals did eat enough and died of 1080 poisoning.

Grain poisoned with 1080 was used to kill rodents in some areas. The 1080 persisted within their bodies, and when these were eaten by scavengers, they sometimes became part of a fatal food chain.

A variety of birds and mammals ingested enough 1080 to kill themselves. These included dogs, foxes, raccoons, badgers, hawks, and eagles. The deaths of these "nontarget" species helped to arouse public opinion against predator control. There were also objections to the inhumanity of the methods—to the fact that predators often spent a day or more with a foot held in a steel trap or died in violent convulsions from 1080. And why, some people asked, were public funds spent to kill wildlife on public lands to benefit private industry? To some people, the widespread use of poisons indicated that the war on predators was nearing an end, with a human "victory." Many citizens, especially in the eastern United States, began to protest the "poisoning of the West."

A great controversy developed. On one side were those who still believed that predators were "bad" or that they caused great economic damage and needed more, not less, control. On the other side were increasing numbers of people who believed that "good" and "bad" labels for animals were naive and who called for a truce in the war. Emotions rose on both sides and were expressed in public meetings, legislative sessions, and letters and telegrams sent to government officials. And at the center of the storm was, of course, the coyote.

In 1963 Secretary of the Interior Stewart Udall appointed a committee to evaluate the federal predator-control program. All five members were wildlife experts, and the chairman was Dr. A. Starker Leopold, a professor of wildlife management and the son of the late Aldo Leopold. The Leopold Committee, as it was called, presented its report in 1964.

The committee stated that all wildlife has value and a place in nature and that large predators are "objects of fascination to most Americans, and for every person whose sheep may be molested by a coyote there are perhaps a thousand others who would thrill to hear a coyote chorus in the night."

According to the Leopold Committee, the Division of Predator and Rodent Control was killing many animals unnecessarily. Instead of setting a good example for the proper scientific management of all wildlife, it was promoting a naive and outmoded attitude about predators. The committee suggested ways to reform predator-control efforts. These included a reduction in the program, tough controls on the poison 1080, and research aimed at finding ways to reduce livestock losses without killing so many predators.

Soon the Division of Predator and Rodent Control was given a new name: Division of Wildlife Services. There was considerable change in personnel, rules, and regulations in the Washington, D.C., headquarters of the agency. However, making changes out on the vast Western rangelands was another matter. The several hundred men who were paid to trap and poison were given a new title ("district field assistant"), but most of them were the same people. Many had spent decades killing predators, justifying their work, and encouraging others to share their views. New guidelines from Washington weren't likely to change their ways. A field agent in Wyoming said: "They notified us that we weren't supposed to go out and kill every coyote any more, just the ones that were doing the damage. But they didn't tell us *how* to do that. So we kept on doing things in exactly the same way, and nobody seemed to care."

Besides, livestock associations and individual ranchers contributed a sizable part of the money spent on predator control. This fact helped keep the focus on predator killing, with little regard for the desires of people who worried about the environmental effects of predator control.

Name changes, new guidelines, and good intentions were not enough to alter significantly the course of the war on predators. Poisons were still used indiscriminately. For example, federal guidelines called for just one 1080-

poisoned carcass per township (about thirty-six square miles). This was considered adequate to kill many coyotes because they often travel several miles in search of food. Other predators, such as foxes and badgers, have much smaller home ranges, so few of them would encounter the poisoned meat. In actual practice, however, there were sometimes several 1080-poison baits per township, placed there by field agents disobeying the rules of the Division of Wildlife Services or by ranchers using poison illegally.

Abuses like these brought further protests from environmentalists. In 1971 a seven-member Advisory Committee on Predator Control was appointed and directed to make a new study. The chairman was Stanley A. Cain, former Assistant Secretary of the Interior. The Cain Committee made its report in late 1971.

Some of its findings were similar to those of the Leopold Committee. For example, the committee's report stated that "today's society places as high a value on prairie dogs, eagles, and coyotes as does the grazing lessee on public lands or the owner of a ranch on his flocks of sheep." The report called for major changes in the predator-control program, including this one: "The foremost need is to discontinue all use of toxic chemicals in predator control. . . . We believe that all necessary killing of coyotes or other predators can be accomplished by means other than use of poison baits."

Although the Cain Committee made fourteen other recommendations, this one brought quick results—and more controversy. Early in 1972 a presidential order banned use of predator poisons on all public lands, including national forests, and also on rangelands where grazing is administered by the Bureau of Land Management. Then the Environmental Protection Agency barred interstate shipment

The predatory coyote has become a national environmental issue, resulting in articles in newspapers and magazines, as well as in publications of environmentalists and sheep raisers.   (*Laurence Pringle*)

of such poisons as 1080. This was aimed at keeping animal poisons out of the hands of ranchers. (Nevertheless, a "black market" in poisons developed, and some illegal poisoning continued, even on public lands.)

With their favorite poisons now illegal, sheep raisers

protested to their political leaders. Twenty-one senators from Western states signed a letter that accused the Department of the Interior of failing to protect livestock. Nearly all governors of Western states urged the reinstatement of predator poisons. Great pressure was applied in Washington, both on Congress and on the White House.

The predator-control controversy reached a new emotional peak, with some sheepmen saying that the very survival of their industry was threatened by increasing numbers of coyotes. At the other extreme were some environmentalists who claimed that the ecological health of the land was at stake. They said, "Coyotes keep populations of rodents and rabbits in check, and we must save them from extinction."

The truth lay somewhere in between, and one recommendation of the Cain Committee was aimed at finding that truth. In its 207-page report, the committee pointed out, again and again, that there was a huge lack of solid information about predators and especially about coyotes and their effects on sheep. It was remarkable, the committee concluded, that the predator-control program had "flourished over the decades without the objective information to warrant it, even in the face of public criticism."

The committee recommended a greatly increased research program, not on new and better poisoning methods, but on such basic questions as the actual losses of sheep to coyotes. Federal funds were appropriated for research, and many studies were begun. At long last it seemed that we would learn the truth about real sheep and real coyotes.

TWO

Yes, Coyotes
Kill Sheep

In 1541 Francisco Coronado led an expedition into the heart of North America, seeking gold. He found none and retreated to Mexico, leaving behind some Spanish padres and their flocks of sheep. A century later, settlers on the East Coast of North America were also raising sheep. As a new nation developed and flourished, so did the sheep industry.

Until late in the nineteenth century, sheep outnumbered cattle in the United States. The sheep-raising industry had its troubles, especially in battles with cattle ranchers over public grazing land in the West. But it continued to grow. During the Depression of the 1930s, the federal government did everything possible to stimulate agricultural production. As a result, sheep numbers reached a peak of more than fifty million head early in 1942.

Costs rose sharply during World War II. As ranchers' profits dropped, so did the numbers of sheep. The industry's wool faced new competition in the form of synthetic fibers such as nylon and rayon. Many experienced shepherds, wranglers, shearers, and other workers joined the armed forces or worked in factories; after the war few of them returned to their former jobs. Today labor continues to be a problem for the sheep industry, even though Congress allows the immigration of Basque shepherds from Europe to fill part of the gap. The federal government also subsidizes the sheep industry through tariffs on wool, lamb, and mutton; incentive payments to wool producers; low-cost grazing leases on public lands; and, of course, financial support and administration of the predator-control program. Despite this help, the sheep industry has been in decline since World War II.

Growing concern about the environment caused part of this decline. Many Western sheep spend at least part of each year grazing on public land. For decades conservationists have complained that this land was terribly overgrazed. Sheep have been called "hoofed locusts" and "arch-predators of plants and soil." If allowed to, they eat plants right down to ground level, then eat part of the roots, leaving a wasteland.

Overgrazing occurred on millions of acres of federal land. Belatedly, the Forest Service and the Bureau of Land Management began reducing the grazing allotments on the lands they administered. This caused further declines in the number of sheep raised.

The sheep industry has also been affected by changing food tastes. People now eat much more beef than lamb or mutton. Many ranchers have switched to the usually more profitable cattle business. By 1972 there were 118 million head of cattle in the United States and only 18.5 million sheep. In 1974 sheep numbers had dropped further to 14.5 million.

Some environmentalists believe that grazing allotments on public lands should be reduced even further, to halt erosion and other effects of overgrazing. This would probably reduce the nation's sheep population still further. However, there are sound environmental reasons for allowing

Damage from overgrazing has been reduced, but not eliminated, on public lands. (*U.S. Bureau of Land Management*)

sheep raising to continue as long as overgrazing is avoided.

Unlike cattle, sheep produce two crops: wool and meat. Wool accounts for only about a third of American sheep raisers' income; in fact, most of it is sold in foreign markets, not in the United States. But the value of wool may rise. It is a natural fiber. Dacron, orlon, and other synthetic fibers are made from petrochemicals, which will become increasingly expensive as the earth's supplies of petroleum are used up.

Changing patterns of food preferences may also increase the value of sheep. Experts on nutrition believe that Americans eat too much meat, especially fat-rich beef. People may eventually reduce the meat in their diets and may also shift to meat from sheep, hogs, and chickens.

A line of sheep shearers gathering up wool. Sheep raising is likely to continue being an important industry in the West and may grow in importance. (*U.S. Department of Agriculture*)

These animals convert plant material to protein more efficiently than cattle do, with less food energy lost. Besides, sheep can be raised in sagebrush and in steep terrain where cattle can't go and in areas with very little rainfall. So it seems likely that the sheep industry will survive and may become more important to the nation, and perhaps to the world.

The small flocks raised in the East and Midwest are usually a minor part of a farmer's business, while Western ranchers often depend on sheep for their entire livelihood. Large flocks are frequently kept on fenced-in areas of the Great Plains and the Southwest. Elsewhere, in much of the West, sheep follow the grass. Flocks of 1,800 to 3,000 head are herded, or trucked, to mountain grazing areas in the summer, then brought back to the lowlands for the winter. Most of the lambs are shipped off to feedlots or slaughterhouses at summer's end; at no more than six months old, they weigh between 95 and 105 pounds.

No matter where sheep are raised, some die during the course of a year. Many losses occur on the lambing grounds. Some ewes die while giving birth. About 5 percent of lambs are born dead. Other lambs die from birth defects, from inadequate milk, and from abandonment. Sheep also die as a result of storms. In the spring of 1973, Western blizzards destroyed entire flocks of pregnant ewes.

Losses continue when sheep are on their summer ranges, as animals of all ages have accidents, eat poisonous plants, get lost, or are accidentally left behind when the main herd is moved. Throughout the year, but especially while lambs are young, there are also losses to predators.

Total losses can be reduced if lambing is done under the shelter of sheds, but this is costly and is not possible on the open range. Western sheepmen have come to expect losses and conclude that they can't do much about most of them. They also feel powerless to do much about their labor problems, or about the market price of wool or lamb. But predators—now there's a nearby problem about which something can be done.

Some sheepmen have a deep hatred for coyotes and commit acts of unspeakable cruelty on coyotes found alive in traps. They feel quite justified, perhaps because they've seen a coyote kill a lamb even as it was being born, or because coyotes sometimes kill several ready-for-market lambs, then run off without eating a bite. It probably seems to them that coyotes are cruel animals that kill for fun.

As far as we know, humans are the only animals that

are consciously, deliberately cruel. Sometimes coyotes kill because they are hungry. At other times they attack because a prey animal flees or exhibits some other behavior that triggers an automatic response. Sheep are rather flighty, panicky creatures, so they may inadvertently send "attack me" messages to coyotes. The unfortunate result is dead sheep that aren't used by either coyotes or people.

WHAT ARE THE REAL LOSSES?

Some sheepmen tend to exaggerate their losses to predators. A careless shepherd may also blame coyotes for losses caused by his own poor work. In some instances sheep raisers have been urged to exaggerate predator losses by employees of the Division of Wildlife Services, in a misguided effort to maintain their own jobs. Because of all these factors, predator losses reported by the sheep industry have always been viewed with skepticism by environmentalists.

Over the past fifty years there have been several studies aimed at discovering the *real* losses to predators. Some were based on questionnaires sent to sheepmen. Usually there was no checkup on the reported figures, so the reliability of these studies is sometimes doubtful. Perhaps the most reliable study of this kind was conducted in Utah during 1968 and 1969. The investigators personally met and interviewed about 20 percent of all Utah sheep raisers. The ranchers were asked to estimate their total losses, and also their losses to predators. They reported that about 3 percent of their sheep had been killed by predators, mostly coyotes.

This may seem to be a rather small and acceptable loss. But the average profit of Western sheep ranchers is quite low, reportedly about 2 to 4 percent of their investment. Fewer losses to predators could increase profits significantly. Also, the Utah study showed that some ranchers sustained losses much higher than the average. For them the losses to predators were a serious blow. Furthermore, the study was conducted during a time when Utah's coyote population was lower than usual. This suggests that the average losses might be higher when coyote numbers were greater.

The 1971 report of the Cain Committee stressed the need for new, unbiased studies of sheep losses. A number of investigations have been completed or are in progress, but several years may pass before a clear picture of predator losses emerges. The Fish and Wildlife Service wants to determine losses throughout the full range of Western sheep ranching, including plains, mountains, open ranges, and fenced-in areas. There will be comparisons between areas where there is no coyote control and other regions where there is intense control. The flocks under study include those on efficient, well-managed ranches and those on poorly run operations (which often have the highest losses).

In New Mexico, hundreds of sheep in a flock were fitted with collars containing miniature radio transmitters. If a sheep died, its radio soon began sending a signal, which alerted biologists so that they could find the dead animal and determine how it had died. The radio collars enabled them to find sheep that otherwise couldn't have been located or that might have been found so late that decay would have destroyed evidence of the cause of death. Daily searches and observations of vultures in the sky also led investigators to dead sheep. Between March 31 and October 19, 1974, this study revealed that predators (mostly coyotes and bobcats) had killed no adult sheep but had taken 50 out of 271 lambs. This was 15.6 percent of the lamb crop, but a much smaller part of the entire flock, which included many ewes. In 1975, predators killed 12.1 percent of the lambs, plus a few ewes. These losses occurred on a ranch where no predator control was practiced. However, rather intense control efforts were made by ranchers on adjoining land, especially in 1975.

Individual radio transmitters for sheep are costly, so most studies of sheep predation simply involve close observation of a flock, day after day, as it goes through the summer and early fall. Investigators are trained to examine dead sheep in order to find the cause of death. They skin each carcass. Signs of heavy bleeding around wounds indicate that the sheep's heart was beating when it died and is evidence of a predator kill. If there are toothmarks or other wounds but no sign of heavy bleeding, the animal probably died of natural causes and was fed upon after death.

The location and type of wound indicate whether a sheep was killed by a coyote, bobcat, eagle, dog, or some other predator. Nine times out of ten, coyotes attack a sheep's throat. Dogs bite all over a sheep's body. Predation studies have revealed that dogs sometimes cause serious sheep losses. In parts of New Mexico, free-roaming dogs did more damage to sheep flocks than did wild predators. In two California counties, dogs were responsible for 29 percent of the losses to predators, coyotes for 66 percent.

Even though great care is taken in the studies of sheep losses, it is virtually impossible to account for all the animals that die. Some sheep just disappear. Biologists in California divided losses into three categories: natural, predator, and "thin air."

When lambs are quite small, coyotes can carry them

A coyote can kill lambs and even full-grown ewes rather easily and often eats little of its prey. *(U.S. Fish and Wildlife Service)*

off without leaving a trace. A Montana sheepman, Bill Stockton, described one such case: "One spring I was losing lambs to a pair of coyotes at the rate of seven or eight per week. We were unable to backtrack them, and a thorough search of their usual denning areas failed to yield any clues as to their whereabouts. Finally, with some luck, the government trapper located a den over five miles from my band and killed the dog [male] and all the pups. The killing among my lambs immediately stopped.

"As it turned out, this pair, singularly or together, had been crossing a creek, a highway, two woven wire fences, a county road, a railroad, and another creek to kill from my band, when all the time another band of sheep was grazing within a mile of their den site. Not only that—they had carried the lambs that distance."

Within a few weeks after birth, lambs grow too heavy to be carried off by coyotes, so this ceases to be a problem for those who assess losses to predators. There are other problems, though. Investigators suspect that some of their conclusions about cause of death may be wrong, even when a sheep is found and examined. This problem was explained by Dr. James R. Tigner of the Fish and Wildlife Service, who conducted a study of sheep predation in Wyoming.

Dr. Tigner said, "Suppose a ewe has twin lambs. A coyote attack on its flock, or just the approach of a coyote, may cause the ewe to abandon one lamb. Later the lamb is found dead with no milk in its stomach. The official cause of death, based on what we can observe, is starvation. The real cause is coyote. On the other hand, coyotes undoubtedly kill some sheep that are sick or injured—deaths that would occur anyway."

Coyotes do kill *some* sheep that would have died of other causes. However, in defending coyotes, some people claim that predators actually aid ranchers by killing *only* sick, weak, or injured sheep. So far, there is no evidence to support this idea. There *is* evidence that it is true of predator-prey relationships that have evolved over millions of years—between coyotes and jackrabbits, for instance. But the match between a coyote and a domesticated animal is quite different.

In the opinion of Dr. Tigner, "A coyote doesn't need one-tenth of its predatory ability to kill a sheep. Sheepmen claim that coyotes often take the biggest and best lambs. They're probably right. Sick and weak lambs may stay in the center of a flock, with healthier animals on the fringe. By virtue of their vigor they become targets for predators."

There is still a lot to learn about sheep and their relationships with coyotes. Do certain sheep inadvertently "select" themselves to be prey by their behavior? Is there an ideal flock size that enables sheep to resist predation? Questions like these may be answered by observations of coyotes and sheep kept in large open pens. One such study, conducted in California, was reported in the July 1976 issue of the *Journal of Wildlife Management*.

Wildlife biologists watched through a one-way glass window as coyotes and sheep were allowed to interact in a sixteen-acre pen. They noticed that fleeing sheep were always pursued, while sheep that did not run were rarely attacked. Defensive behavior by sheep often stopped attacks, too. Lambs and ewes stamped their front feet at approaching coyotes. Rams and some ewes lowered their heads and often charged and chased coyotes. One ram butted a coyote with his head, knocking it off its feet. The presence of a ram in a flock seemed to reduce the predation on ewes. However, the investigators concluded that the sheep in the pens showed defensive behavior mostly be-

Studies at the Hopland Field Station, University of California, showed that ewes can sometimes defend themselves well against coyotes. The photo below shows a captive coyote gripping a 55-pound lamb by the throat.   (*Guy Connolly*)

cause the surrounding fences kept them from fleeing. On the open range they would be more likely to run than to take a stand.

The biologists were able to observe the sheep-killing behavior of coyotes. After a coyote first bit a sheep, the killing time averaged thirteen minutes. Each coyote ran alongside the fleeing sheep, then clamped its jaws on its neck just behind the ears and braced its feet to stop the sheep. Then the coyote shifted its grip toward the underside of the throat. Whether the sheep ran or not, the coyote simply held on, sometimes for twenty minutes. Eventually the sheep suffocated.

Some similar findings were made in the first year of a long-term study conducted on the 8,383-acre Eight Mile Ranch of western Montana. During the first stage (1974–1975) the investigators were Donald Henne, a graduate student at the University of Montana, and his assistants. The ranch had lost many sheep during 1973 despite the death of at least thirty-seven coyotes. But there was little predator control in the general area. Also, the ranch's pastures were fairly small and open, providing good conditions for finding dead sheep. The ranch owner agreed to cooperate in the study; he was paid for all predator losses by the Denver Wildlife Research Center of the Fish and Wildlife Service. The stage was set; in March 1974 Donald Henne began the first comprehensive study of sheep predation on a ranch while there was *no predator control.*

THE EIGHT MILE RANCH STUDY

No coyotes were killed for seven months, beginning March 15, 1974. A few were shot or trapped in the fall, but this occurred after the lambs had been shipped to market.

Henne was trained by veterinarians to recognize sheep diseases as well as different kinds of predator wounds. His observations of losses began in the March lambing period.

Lambs were born and cared for in sheds or in small corrals attached to sheds. Both lambs and ewes received good care during the lambing period. Nevertheless, about 9 percent of the lambs died then. The natural deaths continued after the ewes and lambs were moved out onto the range. Of the total herd in pastures (about 2,000 head), about 3 percent died between March 1974 and March 1975. Henne examined as many of these natural deaths as possible; the most common cause of death was pneumonia.

During the same period, predators killed 16.9 percent of the sheep on the ranch—355 lambs and 94 ewes. Coyotes killed all but a few, and only twelve sheep of the original herd were unaccounted for.

These are heavy losses. Henne was also able to gather other valuable information about the coyote-sheep interactions on the ranch. About half of the coyote kills were fresh enough so that he was able to determine whether the sheep were healthy before attack. The results of these examinations were compared with fourteen lambs shot at random. In both groups, about three-fourths of the lambs were healthy. These figures indicated that the coyotes were not "culling" the herd by taking mostly unhealthy sheep. This finding was supported by direct observations of Henne and his assistants. They saw sheep that were sick, wounded, or limping and noted their tag numbers in case the animals were later found dead. Two ewes were severely crippled. Yet most of these sheep lived for several weeks, and some survived the entire year.

Henne suspected that these unfit sheep survived because they were less active than healthy sheep. An active sheep is more likely to be noticed, and perhaps, as noted earlier, its liveliness stimulates a coyote attack. Coyotes killed more female than male lambs. Henne speculated, "If, for some reason, ewe lambs were detected more easily or broke from the main group more often when under attack, their chances of being killed would be increased."

Coyotes usually attacked just before dawn, while herds were on their bedding grounds. Many of the kills were made in ditches, ravines, and stream bottoms—places where a fleeing sheep would be likely to slow down or lose its footing and be more easily caught.

One of Henne's assistants witnessed an early-morning kill. Two coyotes entered a 193-acre flat pasture where sheep were kept. Only one coyote, the male, attacked. It trotted toward the flock. The sheep bunched and ran as the coyote trotted in ever-smaller circles around them. One 33-pound lamb broke from the flock, and the coyote quickly "herded" it farther away. Then the coyote bit the lamb on the neck. The lamb kept moving and stayed on its feet for three minutes, but the coyote never lost its grip. The lamb fell, rose again, and ran, but the coyote hung on, and the lamb soon fell for the last time.

About 72 percent of the coyote-killed sheep had neck or throat wounds. Henne's observations about the actual cause of death agreed with those from the pen studies in California. The powerful jaws of coyotes blocked air passage to the lungs so that sheep died by suffocation rather than loss of blood. Sometimes small lambs were gripped by the head and their skulls crushed.

During the study, many sheep were wounded, some so severely that they were beyond help and were killed by Henne or his assistants. About half of the coyote-killed sheep were eaten lightly or not at all. The coyotes seldom returned to a kill. Many carcasses were "reduced to mere skin and bones within hours by ravens, magpies, and golden eagles." Besides, the coyotes had little trouble getting fresh meat.

A follow-up study of predation of the Eight Mile Ranch was conducted from March 1975 to March 1976. The flock's size was increased by more than a thousand sheep, but the predation toll was about the same: 16.3 percent.

The predator losses at Eight Mile Ranch were much higher than average for Western sheep ranches. Are these the kind of losses we could expect if coyote control efforts were stopped? Not necessarily. Though well managed, the ranch may be (for some unknown reason) especially vulnerable to coyote predation. Further studies at this ranch and at others may help clarify the matter. However, the predation loss of more than 16 percent was real and well documented. It should be sobering information for those people who advocate an end to all coyote control efforts.

Sheep ranchers are not alone in their war on coyotes. In Texas, they are joined by goat ranchers. The predators kill goats as easily as sheep, and 90 percent of the nation's goats (nearly two million head) are raised in Texas. In 1973, a spokesman for the Texas Sheep and Goat Raisers' Association said, "Texas cannot maintain its sheep and goat industry without the use of chemical poisons to help control the predators."

Cattle are another matter. As a rule, most cattle ranchers bear no grudge against coyotes. Healthy adult cows or steers are too large and powerful for coyotes to kill. The 1964 Leopold Committee report concluded, "In great areas of the West, cattle and coyotes seem to live amicably together, with no reported losses whatsoever. On rangelands occupied by cattle, and not used by sheep . . . there is little justification for general coyote control."

Some cattlemen oppose coyote killing and forbid hunting and trapping on their land. As a general rule, however, poisons, traps, and other control efforts have been used on cattle ranches and on public lands where cattle graze.

During the 1970s there were increasing reports of coyote predation on calves. Whether this was simply a result of increased numbers of cattle is not clear. Some ranchers who formerly raised steers had changed to cows and calves; as a result, they may have experienced coyote predation for the first time. In 1973 a representative for the Texas and Southwestern Cattle Raisers Association stated to a Congressional agricultural committee, "Predators have become a cause for concern in every section of our state, for producers of all kinds of livestock."

In addition to sheep and goat ranchers and increasing numbers of cattlemen, one other Western agricultural group complains about coyotes: farmers who raise cantaloupes and other melons. Yes, coyotes prey on melons. Like foxes and many other predators, coyotes are *omnivores*—they eat both animal and plant food, including apples, berries, and whatever else they find. Sometimes a coyote trots through a melon field, takes a bite or two from many ripening fruits, and causes considerable damage. Coyotes also strip fruit from the lower branches of trees in orange, avocado, and plum orchards.

It is Western sheep growers, however, who are the main advocates of continued killing of coyotes as long as this seems to be the only way to reduce losses to predators. In 1974 industry spokesmen claimed that coyotes had killed nearly 800,000 sheep ("mostly young lambs") worth nearly seventeen million dollars. Some observers believed these

figures to be double or triple the real losses. However, the long-overdue studies of predation on sheep have left no doubt: coyotes do kill sheep and can be a serious threat to the livelihood of some ranchers.

In Texas, goat ranchers join sheep ranchers in complaints about coyote predation on their livestock. (*U.S. Department of Agriculture*)

THREE

Getting to Know
the Real Coyote

Although humans and coyotes have shared the same environment for centuries, most observations of coyotes have been made over the sights of rifles. Fortunately this situation has changed during the past few years. Wildlife biologists have made several studies of coyotes in the wild, and further studies are planned. In light of the results so far, both sheepmen and environmentalists have had to give up some cherished notions. The mythical coyote dies as a picture of the real coyote emerges.

The coyote is related to the wolf, but it differs in several ways, some of which have enabled it to thrive while the wolf has become nearly extinct in the original forty-eight states. The coyote is not a particularly big animal. Coyotes in the West usually weigh between twenty and thirty pounds and stand about one-and-a-half feet high. This size is an advantage in survival, not just because a coyote can hide in rather small places, but because its food needs are not as great as those of larger predators. A coyote can—and frequently does—support itself on meals of mice, grasshoppers, or other small prey.

Coyotes sometimes live in family groups, or packs. These groups are usually not as large or as long-lasting as the close-knit social packs of wolves. However, very little is known about the social lives of coyotes. Adult females seem to stay in a territory of a few square miles. Adult males wander over much larger areas, as do young coyotes of both sexes. Some coyotes that were trapped, marked with identity tags, and released were later found as far as 180 miles from their starting point.

Coyotes often hunt alone or in pairs. A cooperative team sometimes catches prey in situations where a lone coyote would fail. One animal may dig into a ground squirrel's burrow while the second coyote waits for the squirrel to flee from another hole. Coyotes have also been observed using teamwork in order to outwit or exhaust other prey. Sometimes one coyote will make a display of jumping, whirling about, and acting "crazy," drawing the attention of some prey animal while it is being stalked by a second coyote.

The coyote has excellent hearing and sense of smell. It can hear the rustlings or squeakings of meadow mice in their runways beneath the snow and detect the faint scent of a hunter a mile or more away. It has been called "the clever coyote"; so far, its intelligence has helped it survive the best efforts and weapons that people can muster against it.

The exact original range of the coyote is unknown, but its present range is probably three or four times larger than it was when North America was colonized. Coyotes now live from Costa Rica and Honduras in Central America to far northern Alaska, and from California to New England and Georgia. They were unknown in eastern North America when it was first settled. The death of wolves, bears, and cougars left foxes and bobcats the largest predators in most of the remaining habitat. During the past few decades,

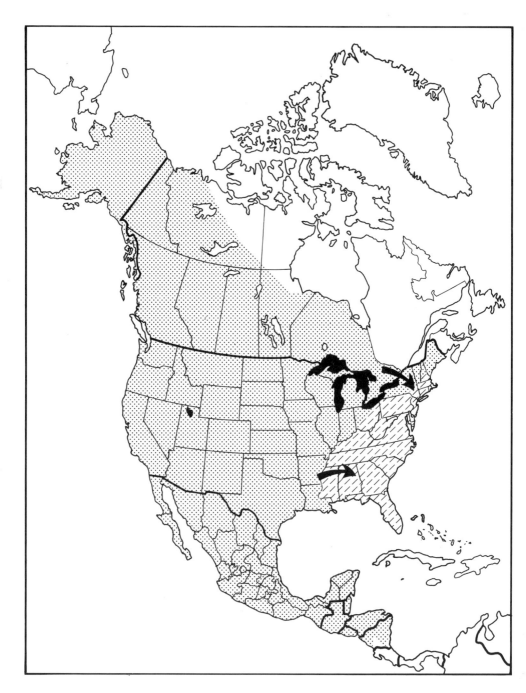

By the mid-1970s, the coyote's range (*dotted area*) covered nearly all of North America. There were well-established populations in New England and in the Adirondack Mountains of New York State and reports that coyotes were spreading into other Eastern states (*lined area*).

however, a northern subspecies of the coyote (possibly the result of some cross-breeding between coyotes and wolves or coyotes and dogs) has spread through Canada into the northeastern United States. Several thousand of these coyotes, some weighing up to sixty pounds, now live in New York State and the New England states.

Although all coyotes are of the same species, *Canis latrans,* there are nearly twenty subspecies, which differ slightly from one another in appearance, size, and other characteristics. Coyotes of the Southwestern deserts are smallest and lightest in color; those of cold Northern forests are largest and darkest.

In rural parts of eastern North America, people are sometimes startled to hear the lonely wails of coyotes. Sometimes called the song dog, the coyote may be the most musical of all land mammals. A four-year study of coyote sounds in Wyoming revealed that they have a "language" of at least eleven different calls. The investigator was Dr. Philip N. Lehner, a biologist at Colorado State University in Fort Collins. He gave the coyote sounds these names: growl, huff, whine, yelp, woof, bark, bark-yip, lone howl, group howl, group howl-yip, and the greeting song. Each sound seemed to have a different meaning to coyotes, though each is part of series that coyotes can make which express different degrees of "feeling." (Coyotes also communicate through scents and body postures.)

Coyotes give lone howls when they are separated from other members of their pack. When a low-ranking pack member is reunited with more dominant pack members, it gives the soft, wavering greeting song. The most spectacular of all coyote sounds—the group howl and group howl-yip—often last for three minutes. These are believed to strengthen the bonds among pack members and also to declare to other packs that a certain territory is occupied.

Coyote "music" is a symbol of wide-open Western spaces. Coyotes are usually quiet and secretive when close to humans. They live within the borders of Denver, Portland, Los Angeles, and many other cities. Concerts at the Hollywood Bowl in Los Angeles are sometimes accompanied by a coyote chorus. The bowl is in Griffith Park, just six miles from the city's center. The park consists of about six squares miles of wooded ridges and valleys and supports a population of about twenty coyotes.

This coyote, captured in Massachusetts, is typical of the dark-coated variety that is now living in New England. *(Laurence Pringle)*

A coyote howling at its den, with a meal of jackrabbit in view.
(*U.S. Fish and Wildlife Service*)

Although many people in and near Los Angeles like to hear the wild sounds of coyotes, the city's Department of Animal Regulation receives more than a hundred complaints each year about another matter: coyotes killing pet dogs and cats. Dogs and cats that have gone wild are also coyote prey. There is speculation that coyotes don't just kill these animals for food.

In the wild, predatory species compete to some extent with one another for food and other resources. Wolves were once "top dog," and wildlife biologists have seen wolves kill coyotes. Now coyotes are "top dog" in many areas. They have been known to kill foxes, skunks, and other smaller predators. For that matter, they sometimes kill other coyotes. Observations of coyote packs in Wyoming's Grand Teton National Park showed that they sometimes fought over elk carrion in the winter and also killed the pups of neighboring packs. Perhaps a few Los Angeles pets are also victims of the coyote's natural aggression toward other predators.

Coyotes compete for food with foxes, badgers, bobcats, and other predators more directly than the larger wolves ever did. As a result, when the numbers of coyotes are reduced, there is usually an increase in the numbers of other predators. Red foxes became more abundant in the western Dakotas when the coyote population dropped. In Utah, Wyoming, Arizona, and Texas, the numbers of badgers, bobcats, and foxes increased after coyotes were reduced by trapping and poisoning.

The numbers of badgers (*above*) and bobcats (*right*) tend to increase when coyote populations decline.

(*U.S. Fish and Wildlife Service* and *Laurence Pringle*)

Donald S. Balser, who directs research on predator damage at the Denver Wildlife Research Center of the Fish and Wildlife Service, has said, "Some people urge that we 'let nature take its course,' without considering how humans have already altered the environment to the advantage of coyotes. We have unwittingly created a kind of 'predator monotype' in much of the West. For me, there is a special benefit when coyotes are reduced somewhat; I like to see a greater variety of predators on the land."

UPS AND DOWNS OF COYOTE NUMBERS

Regardless of control efforts by people, coyote populations are kept in check by most of the same factors that have operated for thousands of years. Coyotes catch diseases such as mange, distemper, and canine hepatitis. They have accidents. Sometimes they starve to death. Winter can be a grim time, especially in the North. Survival may depend on finding the carcass of an elk or some other carrion. In the province of Alberta, Canada, a biologist found that coyotes relied on carrion for 70 percent of their winter food.

Coyote pups are especially vulnerable to diseases and accidents and are also eaten by great horned owls and golden eagles. Adults are sometimes killed or wounded by the slashing hoofs of deer, elk, and pronghorn antelope.

We don't know much about this natural mortality of coyotes. More is known about "unnatural mortality"— coyotes killed by people. More than 81,000 coyotes were killed in fiscal year 1975 (from July 1, 1974, to June 30, 1975) by the federal Animal Damage Control Program. (The Division of Wildlife Services was reorganized in mid-1974 and renamed the Animal Damage Control Program.) Most of these deaths occurred on or near sheep ranges,

Hunters kill many thousands of coyotes each year. However, wildlife biologists believe that such natural factors as disease and food supply are the chief controls of coyote numbers over most of their range.

(*U.S. Fish and Wildlife Service*)

which take up about 25 percent of the West. Throughout the West, however, many more thousands are trapped or shot by private citizens. Few hunters deliberately set out after coyotes, but most will shoot when they see one. Some states and counties pay bounties for dead coyotes. According to a 1974 study conducted in California, hunters annually shoot more than 80,000 coyotes in that state alone.

How can the coyote survive these losses? Rather easily, it seems. There are an estimated one million coyotes in the West. Females have four or more pups in a litter.

The species has the potential of tripling its numbers each year. Also, coyote reproduction is "density dependent." If, for example, a coyote population in a Western valley is using all the available food and other resources, its numbers, or density within that area, cannot increase. The coyotes continue to mate and have young, but the new generation is usually just sufficient to replace the animals that die. "Extra" young coyotes leave the valley in search of new homes.

If the population density lessens, however, coyote reproduction changes. Suppose that intense control efforts are aimed at coyotes and the population in the valley drops. Its density is less than before. There is an abundance of food, dens, and other resources—more than enough to go around. This situation can have at least three effects. Other kinds of predators may be better able to survive and increase. Coyotes that immigrate into the valley have greater chances of finding a satisfactory home. And the resident coyote population produces more young than usual. Many yearling females may mate and have pups. All females produce more pups than they do normally.

Field studies revealed how this worked in southern Texas. Coyotes were extremely abundant in some areas, and females had an average of 4.3 pups per litter. In other areas intense control efforts held coyotes to a low population density. The females in these areas had average litters of 6.9 pups. Thus, the coyote-killing efforts had the effect of stimulating reproduction, and the coyote populations were able to survive.

Nearly all animal species (including those that coyotes prey upon) possess this ability to adjust reproduction to environmental conditions. The animals respond automatically, unconsciously, to "feedback" from their surroundings.

Female coyotes produce more pups than usual when the level of a coyote population is reduced by poisons or other control efforts.

(*U.S. Fish and Wildlife Service*)

Feedback from the environment may suppress reproduction as well as increase it. A declining food supply may be a "signal" for coyotes to produce fewer young.

More than ten years of study in the Curlew Valley of Utah and Idaho has shown how changes in food supply can affect the numbers of coyotes. The Curlew Valley is a cold northern desert, with less variety and abundance of plants and animals than southern areas with more warmth and rainfall. Black-tailed jackrabbits are common, and the remains of these animals show up frequently in the stomachs of coyotes. Year round, Curlew Valley coyotes depend on jackrabbits for three quarters of their food.

In 1962 the jackrabbit population began to drop. A year later the coyote population also declined. Rabbit numbers kept dropping, and the coyote population followed. Then, in 1967, the numbers of rabbits began to rise. A year later the coyotes began increasing too.

The results of this and other similar studies upset a cherished myth held by many people. Those who defend predators often do so by emphasizing their economic value (because others emphasize the economic harm they do). Someone once estimated that a coyote saved a farmer or rancher $21 a year by killing rabbits and rodents. Coyotes do eat lots of rabbits and rodents; from this fact some people leap to the conclusion that coyotes actually control populations of these plant eaters. Therefore, it seemed to follow that an absence of coyotes would spell doom for crops and other plants as the numbers of rabbits and rodents rose astronomically, out of "control."

The idea that coyotes control rabbits and rodents seems to be wishful thinking. The Curlew Valley study showed that coyote numbers rose or fell according to the population of their basic food, the jackrabbit. The coyotes did not control the rabbits; instead, the rabbits appeared to play a large role in determining the coyote population.

Coyotes in the Curlew Valley had some effect on rabbit numbers. When the rabbit population was low, coyote predation drove it even lower. The result was the extension of the period of low rabbit density. Nevertheless, coyotes are only one predator of rabbits, and predators are only one factor that keeps rabbit numbers in check. If all coyotes were wiped out, other predators, plus diseases and other factors, would normally keep rabbit numbers from soaring.

Studies by ecologists have shown this same effect in all sorts of plant-animal communities. Many factors com-

Jackrabbits have regular ups and downs in their populations, and these cycles seem to affect the numbers of their main predator, the coyote. *(U.S. Fish and Wildlife Service)*

bine to determine the population level of a species. As a rule, predators *do not* control populations of plant eaters, even though they may kill many of them. The main exceptions are found in relationships between large plant-eating animals and large predators. Under some circumstances, large predators do seem to have a controlling effect on plant eaters such as deer, moose, and antelope. Biologists noticed that antelope populations rose in some regions during the time poison-bait stations were widely used against coyotes. However, the effect of large predators on populations of plant eaters is often exaggerated. In the case of the Kaibab National Forest deer herd, already mentioned, it is now recognized that an increase in food supply played a large part in the population "explosion." It was not simply the result of less predation.

Studies of coyotes and their prey in Curlew Valley are continuing. However, it is already clear from other studies that rabbits and rodents—the main food of coyotes—have natural ups and downs in their populations. These variations cause similar changes in coyote numbers, which occur regardless of human control efforts.

During the 1970s sheepmen in some states complained that the ban on poisons had caused coyote populations to grow. This may have been true in some regions. However, biologists from the Fish and Wildlife Service knew that coyote numbers in some states were rising as part of their normal pattern, even before poisons were outlawed.

COUNTING COYOTES

Beginning in 1972 the Fish and Wildlife Service, in cooperation with other federal and state agencies, began an annual census of predators in the West. The census method is simple. At fifty places regularly spaced along a fifteen-mile roadside route, a three-foot circle of earth or sand is cleared of all debris. A capsule of attractive-smelling bait is placed in the center of each circle. Then these scent stations are checked daily for five days in a row. Any animal visits are revealed by tracks left in the smooth earth or sand. These are noted by investigators, who smooth over the soil each day so that new tracks can be recognized.

In the course of checking a census line, an investiga-

Coyotes kill fawns of pronghorn antelopes and appear to exert a strong control on pronghorn populations in some Western areas.
(*U.S. Fish and Wildlife Service*)

tor's records might look like this: "no visit . . . red fox visit . . . coyote visit." Throughout the West during the same time of the year (early to mid-September), hundreds of investigators record the same sort of information from nearly four hundred census lines in eighteen states.

The census lines were planned and selected to measure the abundance of coyotes, but they also give some information about other common predators. The survey reveals nothing about actual numbers of predators. However, it shows relative abundance from year to year, and population trends. The 1973 survey, for example, showed that coyote populations rose in most states east of the continental divide and declined in several states of the Far West.

People who believe that coyotes are in danger of extinction may be startled to learn that some biologists doubt whether decades of poisoning, trapping, and other control efforts have had much effect on coyote populations over most of the West. Poisons did cut population density by half or more in some areas, especially in the North. The 1080-bait stations and the M-44 devices were most effective in places where winter food scarcity makes coyotes dependent on carrion. However, some people believe that the effectiveness of all coyote-control methods has declined over the years.

If this is true, there are at least two possible explanations. One is obvious: coyotes are intelligent enough to learn from their mistakes or the observed mistakes of fellow coyotes. And coyote parents can teach their young to avoid similar mistakes.

Another, more far-fetched explanation is that coyote-control efforts have produced a kind of "super-coyote." According to this idea, some coyotes are more likely to feed on carrion, while others are more interested in killing prey.

A coyote approaches a scent station (*cleared circle*), one of thousands used in the annual Western predator census.

(*F. R. Henderson*, from *U.S. Fish and Wildlife Service*)

The use of 1080-bait stations and M-44s killed chiefly the individuals that were attracted to carrion. The more predatory coyotes tended to survive, breed, and pass on their characteristics to the next generation. The result was populations of coyotes that were less and less vulnerable to poisons and more and more likely to attack prey, including sheep. So far there is no evidence to support this idea.

Only in the 1970s were comprehensive studies begun to learn about the effectiveness of coyote-control methods that had been practiced for decades. One such investigation was conducted by two biologists from the University of California at Davis. Their study area was Mendocino County, which occupies 3,500 square miles of north coastal California. About half of the county is grazed by livestock.

Coyote control has been practiced there for the past thirty years and is presently carried out by six full-time hunter-trappers. They used M-44s until 1972, then relied mostly on traps and den hunting.

According to a careful study made in 1973–1974, there are at least four thousand coyotes in Mendocino County. They are least abundant in the heavily forested coastal areas and most abundant in the more open country. The most serious livestock losses occur on ranches that are near chaparral thickets or in steep, wild areas where coyotes can hide and raise their young. The trappers take as many coyotes as possible from these "hot spot" problem ranches and also try to find dens in the nearby wild lands.

From 1970 to 1974 the trappers took an average of 334 coyotes a year. Hunters killed an additional fifty or so annually. The total of this "unnatural" mortality was only a small fraction of the total population. Moreover, the control effort was concentrated in "hot spots." Nothing prevented the rest of the coyote population from "sending in" replacements for those that were killed. According to the biologists who conducted this study, a coyote population can withstand an annual loss of 70 percent of its numbers and still produce enough young to replace them.

Coyotes also have a tremendous ability to bounce back if control is reduced or stopped. Therefore, the past emphasis on numbers of coyotes killed each year is rather meaningless. For decades the Division of Wildlife Services (now the Animal Damage Control Program) pointed to these statistics with pride, while other people came to view them with alarm. In reality, however, these "body counts" have little meaning unless they are related to the number of coyotes that remain.

The same is true of badgers, foxes, and other predators

Dr. Frederick F. Knowlton plays with a captive coyote that is used in behavior studies at Utah State University. (*Laurence Pringle*)

that were sometimes victims of traps and poisons intended for coyotes. Part of the public clamor against predator control was aroused by the discovery of such animals in traps or dead near 1080-bait stations.

This concern is unfounded, according to Dr. Frederick F. Knowlton, leader of the Predator Ecology and Behavior Project (U.S. Fish and Wildlife Service) at Utah State University in Logan. "People fail to recognize the ecological dynamics involved," said Dr. Knowlton. "They got upset when they saw *individual* badgers, foxes, and other non-target species killed by 1080, without considering the effects on whole *populations*. Apparently, the reduction in numbers of coyotes caused an increase in most other species of predators."

Eventually, further studies of coyotes will shed more light on their population dynamics. By observing the lives of coyotes in national parks, biologists will be able to learn more about the reproduction and mortality of populations that are not hunted or trapped. The movements and behavior of these populations will also be studied. Some people suspect that trapping and other control efforts disrupt the normal behavior of coyotes and may even cause more sheep predation. These suspicions cannot be confirmed or denied until we know more about the normal lives of coyotes.

The studies of the coyote so far reveal it to be a superbly adaptable animal, perhaps the most successful large predator on earth. If the federal predator-control program is regarded as a war, humans can claim only partial, local victories, while the coyote continues to reign over the main battlefield.

FOUR

No
Magic Bullet

While some biologists investigate the lives of coyotes, others seek ways to reduce livestock predation without killing predators unnecessarily. The traditional aim of predator-control efforts has been "blanket" control—reducing populations of coyotes over large regions. Some sheepmen are convinced that this is the only method that works.

The emerging picture of coyote population dynamics raises some questions about the wisdom of "blanket" control. Suppressing coyote numbers over a large region also stimulates coyote reproduction throughout that area. It may also teach many coyotes to avoid traps and other control devices. So, even from a sheep raiser's viewpoint, the goal of killing coyotes indiscriminately over a whole county or larger region can be questioned.

One of the first alternatives to traps, bullets, and poisons to be investigated was also a form of "blanket" control, but it involved no direct killing of coyotes. This was an attempt to limit coyote reproduction by use of an anti-fertility chemical in baits scattered over the land. "Birth control" for coyotes received a lot of attention in the 1960s. The proposed method had some appealing characteristics: It seemed more logical to prevent births than to kill existing animals; the need for other control methods would be reduced; and since the baits would not make coyotes ill or kill them, the animals would be less likely to learn to avoid them.

This method was field-tested in the Southwest for five years. Baits containing an anti-fertility chemical were spread over large areas during the peak of the coyote breeding season. They were concentrated in places where coyotes traveled. A female coyote had to eat only half an ounce of bait in order to become infertile for the few weeks that were most critical in her reproductive cycle.

The baits also contained a chemical marker that caused a yellow color to appear in an animal's bones and teeth. Later in the spring, when an intense trapping effort collected many coyotes, this marker enabled investigators to identify the animals that had eaten the baits. By examining the reproductive tracts of female coyotes, they could also tell how the anti-fertility chemical would have affected reproduction had the animals lived.

The results were disappointing. The anti-fertility chemical worked all right, as laboratory tests had indicated it would. But the chemical reached only about one third of the coyotes, and less than one sixth of the females. Many coyotes avoided the baits or did not find them. Although there has since been further research on anti-fertility chemicals, the problem of inducing coyotes to eat baits remains. The idea of coyote "birth control" has been shelved, and investigators have turned to more promising control methods.

CONTROLLING "KILLER" COYOTES

Any coyote can become a sheep killer, but evidence suggests that only a fraction of them, perhaps a small one, reg-

Biologists use collars containing radio transmitters to learn about the home range and movements of coyotes. Such basic information may lead to the development of more acceptable control methods.

(*U.S. Fish and Wildlife Service*)

ularly preys on sheep. The most likely killers are adults that are raising young near the spring and summer range of sheep. Therefore, environmentalists have long urged that control efforts be aimed at these "killer" coyotes, or, better yet, at finding a way to turn coyotes away from sheep.

Coyote pups at the entrance of their den—testimony to the coyote's ability to sustain itself in the face of all sorts of control efforts, including "birth control." *(Joseph Van Wormer)*

Fences are effective on some kinds of ranches. Combined with trapping and other control methods, woven-wire fencing provided fairly good protection from coyotes in Texas for several decades. Now most of the fencing is in poor condition. Large areas must be refenced, but ranchers consider the cost too great. Coyotes dig under fences, but this actually makes other control rather easy, as traps can be set at these entryways. But coyotes have also learned to jump fences up to five feet high. Fencing is impractical on the vast public lands where many sheep graze.

All sorts of repellents have been tried on sheep. Biologists have investigated such substances as skunk odor, cougar urine, and a chemical extracted from the skin of toads. In one case, a repellent mixed with sheep-branding paint was used. Bells were also hung around the necks of several sheep in a band. This combination seemed to deter coyotes for a while. But the coyote is a remarkably adaptable creature and has shown an ability to overcome an initial fear of bells, smells, and other repellents.

An attempt was made to scare coyotes away from sheep with a siren. This procedure worked well for a few days, but soon it seemed to be acting like a dinner signal for coyotes. Repellent sounds, however, are still under investigation. Coyotes have a wider range of hearing than humans. They may be vulnerable to high-pitched tones that could be emitted from devices carried by several sheep in a flock.

Although the entire coyote-sheep controversy is a serious matter, it is refreshing to find that some of the people involved have kept their sense of humor. For example, here is how one biologist described an idea for repelling coyotes from sheep. It was simple: just have sheep carry brightly colored plastic streamers around their necks. The biologist

A tame coyote easily leaps a thirty-two-inch-high fence in California.
(*Guy Connolly*)

claimed that this method was amazingly effective—as long as the sheep carried the streamers *fast enough*.

Sheep ranchers also display ironic humor via the bumper stickers on their pickup trucks: EAT UTAH LAMB;

TEN THOUSAND COYOTES CAN'T BE WRONG; and CONSERVE
COYOTES . . . SHOOT A SHEEPHERDER.

During the mid-1970s two very different potential approaches to the sheep-predation problem received lots of
attention. One was a form of blanket control that would not
kill a single coyote. Instead, it would cause coyotes to become ill at the taste, smell, or sight of sheep. Caged coyotes
at Utah State University were fed hamburger mixed with
capsules of lithium chloride. This chemical compound is an
emetic—it caused the coyotes to throw up their meal. Later
they were offered nontreated hamburger and refused it.
However, they did eat their normal meal of dog food. The
lithium chloride had given them an aversion to hamburger.
This process is called aversive conditioning.

Encouraged by these results, the investigators tested
the coyotes with live prey, including lambs. The coyotes
became accustomed to killing these animals for their food.
Then some lamb meat, mixed with lithium chloride and
wrapped in lamb hide, caused the coyotes to vomit. After
that experience they were reluctant to attack lambs. In one
instance, after conditioning, a hungry coyote threw up after
merely sniffing a lamb that was put into its cage. Unfortu-
nately the results of this study have not been completely
duplicated by other researchers. Nevertheless, these results
raised hopes that whole populations of coyotes could be
conditioned to avoid sheep. Mother coyotes might even
teach their young to avoid sheep too. However, many questions were raised about whether this method would work
out on the range. Would a single unpleasant experience
with a bait of lamb and lithium chloride be enough to keep
a coyote from attacking live sheep? If one dose of an emetic
was not enough, would coyotes simply learn to avoid the
baits and still find live sheep to be tempting prey?

The ironic humor of these bumperstickers belies the strong feelings many sheep ranchers have about coyote predation. (*Laurence Pringle*)

To be successful, an aversive-conditioning method must get all, or nearly all, the coyotes in a sheep-raising area to eat enough lamb mixed with an emetic so that they learn to avoid live sheep. Scattering baits and even sheep carcasses over the range is not likely to accomplish this. Experience has shown that many coyotes don't eat baits, which may prove to be a major obstacle to aversive conditioning unless more attractive baits are developed.

The second approach involved a different method of getting an emetic to coyotes. The compound would be put in a plastic collar worn around a sheep's neck. A few sheep

A coyote in pursuit of a lamb, at California's Hopland Field Station. Observations of captive coyotes and sheep are necessary in order to learn about their behavior and to perfect repellents or other methods that reduce predation.
(*Guy Connolly*)

with these collars would be left in an area where coyotes had been attacking, while the main flock was moved to safety. According to this plan, an attacking coyote would bite at the sheep's neck, as usual, and give itself a dose of the emetic. The result: no more attacks from that coyote.

Field tests of aversive conditioning are under way. People who know coyotes well have adopted a rather cautious wait-and-see attitude toward it. The adaptability of coyotes is sometimes underestimated. They have been known to attack a sheep's belly after finding a "repellent" collar around the animal's neck. The goal of teaching entire coyote populations to avoid sheep seems beyond reach, but the method certainly deserves further study.

Many biologists are more hopeful about another kind of collar for sheep. It would contain a poison. An attacking coyote would strike at a sheep's neck, pierce the plastic collar, get a few drops of the poison, and die. It would have no chance to learn from the experience. Toxic collars may be the only effective way of stopping those crafty sheep killers that have learned to avoid traps, M-44s, and other devices.

The first tests of toxic collars were conducted in Texas, in situations where the usual control methods had failed and coyotes kept killing lambs. In thirteen cases, once toxic collars were used, only one more lamb was killed. In another case, just two more were killed. In these trials, the poison was 1080. Later studies in large pens tested the effectiveness of quick-acting cyanide. The results were disappointing. Coyotes stopped their attacks on lambs before getting a lethal dose of cyanide. They may have been alerted to danger by the powerful smell of this poison.

Nevertheless, optimistic reports about toxic collars led to a change in the federal government's ban on predator poisons. In July 1975 President Gerald Ford ordered that full-scale field tests of toxic collars be conducted for a year. At that time, however, toxic collars were still in the development stage—far from ready for full-scale testing.

The possible use of toxic collars as a means of reducing predation on sheep was supported by many environmentalists. They were concerned, however, when in September 1975 the Environmental Protection Agency allowed the use of cyanide in M-44s. To some people, this seemed to open the door to widespread "poisoning of the West" once more. Nevertheless, M-44s are much more humane than steel traps and are also less likely to kill species other than coyotes. The regulations of the Environmental Protection

Agency were aimed at preventing misuse of poisons, but only time will tell whether the rules will be effectively enforced.

Despite extremists on both sides of the controversy, a reasonable solution to the coyote-sheep problem seems to be evolving. Given enough time and funds, biologists are confident that effective ways can be found to reduce coyote predation on livestock without an all-out war on coyotes. The toxic collar may come as close as any method to being an all-purpose "magic bullet." A yet undiscovered smell or sound that repels coyotes from sheep might also be effective. It seems likely that there will eventually be several solutions, rather then one, simply because livestock raising itself varies so much.

One thing seems certain: as long as sheep and other small livestock are raised in the West, and until better predation-preventing methods are developed, thousands of coyotes must be killed each year in order to minimize predation losses. The general public and environmentalists will have to accept this concept. (In fact, most environmental organizations have always supported this view; the Sierra Club, for example, does not oppose killing "problem" animals.)

Ranchers will also have to accept the idea that they will lose some stock to coyotes. This seems inevitable, even if much better control methods are developed. For example, a flock may sustain losses between the time coyotes first strike and the time a control method (such as the toxic collar) is put to work. Livestock raisers who lease federal lands will have to face the fact that they are paying guests

on land that belongs to everyone and is used for many pur-
poses besides grazing. To help establish this fact, the federal agencies that administer the land should follow a
recommendation of the Cain Committee: to revoke or sus-
pend grazing privileges if laws governing the use of poisons
are violated. Of course, sheep ranchers will be more willing
to obey laws regulating poisons when they feel that new
methods give them comparable or better protection.

Other recommendations of the Cain Committee that
have not been pursued may deserve study and testing. One
is an insurance program for livestock raisers. Another is the
institution of a "trapper-trainer" program wherein expert
trappers show ranch employees how to catch coyotes. This
seems to be an effective means of predator control in the
central and eastern states. It may be worth a trial in the
West, though conditions are different there and trapping is
far from ideal as a method of preventing predation.

For too long, national attention has been focused on
predator control, as though coyotes were the only problems
of the sheep industry. They are not. Sheepmen have been
slow to try new methods. "With the exception of feedlots,"
said a Department of Agriculture economist, "the sheep industry has not changed in a long, long time." There is great
opportunity for improvement: using breeds of sheep that
are best suited for certain habitats and climates; increasing
reproduction; providing greater protection during and just
after the time lambs are born. If millions of dollars had
been spent on these opportunities for change rather than
on destruction of predators, the sheep industry might well
be more profitable today.

There is growing concern about the management of public lands ▶
where sheep are allowed to graze.   (*Bureau of Land Management*)

A truce between coyotes and sheep is impossible, but one between sheep raisers and environmentalists may yet be achieved.

(*Guy Connolly*)

Thus, both sides of the coyote-sheep controversy face change and compromise. The conflict has sometimes been pictured as a battle between good guys and bad guys, or between city slickers and country hicks. Strong feelings have a way of creating simple images of complex situations. As wildlife biologist Durward L. Allen once wrote, "We thought we knew about predation, but in reality we only felt."

People will go on caring deeply about coyotes, sheep, and other living things. However, strong feelings and clear vision are not incompatible. It is possible to love nature or to raise sheep for one's livelihood and also to see reality.

The efforts of wildlife biologists are revealing more and more of the complex realities of the coyote-sheep relationship. We can feel *and* know and cooperate in finding solutions. Some coyote predation is to be expected; so is some coyote killing. But a choice between coyotes and sheep need never be made. We can have both in abundance.

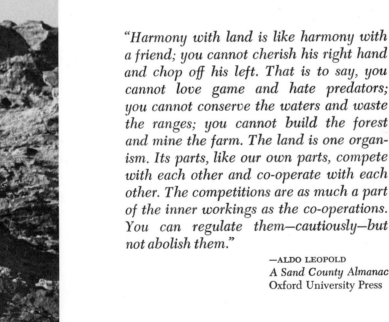

"*Harmony with land is like harmony with a friend; you cannot cherish his right hand and chop off his left. That is to say, you cannot love game and hate predators; you cannot conserve the waters and waste the ranges; you cannot build the forest and mine the farm. The land is one organism. Its parts, like our own parts, compete with each other and co-operate with each other. The competitions are as much a part of the inner workings as the co-operations. You can regulate them—cautiously—but not abolish them.*"

—ALDO LEOPOLD
A Sand County Almanac
Oxford University Press

(*U.S. Fish and Wildlife Service*)

Further Reading

[NOTE: the books and articles listed here represent a full range of published material on the coyote and its predation on sheep. Inclusion of a publication does not mean that the author agrees with its conclusions.]

Beason, S. L. "Selectivity of Predator Control Techniques in South Texas." *Journal of Wildlife Management,* Vol. 38 (1974), pp. 837–844.

Cain, Stanley A., et al. *Report to the Council on Environmental Quality and the Department of the Interior by the Advisory Committee on Predator Control.* January 1972.

Connolly, Guy E., and Longhurst, William M. *The Effects of Control on Coyote Populations.* Bulletin 1872 (1975), University of California Division of Agricultural Sciences at Berkeley.

Connolly, Guy E., et al. "Sheep Killing Behavior of Captive Coyotes." *Journal of Wildlife Management,* Vol. 40, No. 3 (1976), pages 400–407.

DeLorenzo, D. G., and Howard, V. W., Jr. *Evaluation of Sheep Losses on a Range Lambing Operation With-*

out Predator Control in Southeastern New Mexico. Final report to the United States Fish and Wildlife Service, Denver Research Center, 1976.

Fisher, J. "The Plains Dog Moves East." *National Wildlife,* February–March 1975, pp. 14–17.

Flader, Susan L. *Thinking Like a Mountain: Aldo Leopold and the Evolution of an Ecological Attitude Toward Deer, Wolves, and Forests.* Columbia: University of Missouri Press, 1974.

Gustavson, Carl R., and Garcia, John. "Pulling a Gag on the Wily Coyote." *Psychology Today,* August 1974, pp. 68–72.

Henne, Donald R. *Domestic Sheep Mortality on a Western Montana Ranch.* Master's thesis, University of Montana, 1975.

Knowlton, Frederick F. "Coyote Predation as a Factor in Management of Antelope in Fenced Pastures." Proceedings of the Third Biennial Antelope States Workshop, 1968, pp. 65–74.

Knowlton, Frederick F. "Preliminary Interpretations of Coyote Population Mechanics with the Management Implications." *Journal of Wildlife Management,* Vol. 36 (1972), pp. 369–382.

Leopold, Aldo. *A Sand County Almanac.* New York: Oxford University Press, 1966.

Linhart, Samuel, and Robinson, Weldon B. "Some Relative Carnivore Densities in Areas Under Sustained Coyote Control." *Journal of Mammalogy,* November 1972, pp. 880–884.

McMahan, Pamela. "The Victorious Coyote." *Natural History,* January 1975, pp. 42–51.

Munoz, John R. *Causes of Sheep Mortality at the Cook Ranch, Florence, Montana, 1975–76.* Master's thesis, University of Montana, 1976.

Olsen, Jack. *Slaughter the Animals, Poison the Earth*. New York: Simon & Schuster, 1971.

Pearson, Erwin W. "Sheep-Raising in the Seventeen Western States: Population, Distribution, and Trends." *Journal of Range Management*, January 1975, pp. 27–31.

Robinson, Weldon B. "Population Changes of Carnivores in Some Coyote-Control Areas." *Journal of Mammalogy*, November 1961, pp. 510–515.

Ryden, Hope. *God's Dog*. New York: Coward, McCann & Geoghegan, 1975.

Tanner, Ogden. "New England's New Coyote." *Nature/Science Annual*. New York: Time-Life Books, 1974.

Van Wormer, Joe. *The World of the Coyote*. Philadelphia: J. B. Lippincott, 1964.

Wagner, Frederic H. *Coyotes and Sheep: Some Thoughts on Ecology, Economics, and Ethics*. Forty-Fourth Honor Lecture, 1972, Faculty Association, Utah State University, Logan.

Index